GORILLAS

To the trackers and rangers across Africa who find, follow, and help protect wild gorillas. Without their amazing skills and dedication, there would have been nothing to write in this book, and there would be far fewer gorillas in the world.

First published in Great Britain in 2003 by
Colin Baxter Photography Ltd
Grantown-on-Spey
PH26 3NA Scotland

www.worldlifelibrary.co.uk

Text © 2003 Kelly J. Stewart

WorldLife Library Series

A CIP Catalogue record for this book is available from the British Library.

ISBN 1-84107-175-7

Map on page 70 © 2003 Sean Parks and Sandy Harcourt

Photography copyright © 2003 by:

Front cover © Steve Bloom/stevebloom.com
Back cover © K. Ammann
Page 1 © Martin Harvey/NHPA
Page 3 © Andrew Plumptre/Oxford Scientific Films
Page 4 © Martin Harvey/NHPA
Page 6 © K. & K. Ammann
Page 9 © Jorg & Petra Wegner/Bruce Coleman Collection
Page 11 © Steve Turner/Oxford Scientific Films
Page 12 © Daniel Cox/Oxford Scientific Films
Page 15 © Steve Bloom/stevebloom.com
Page 16 © Martin Harvey/NHPA
Page 17 © Michael Nichols/National Geographic
Page 19 © Martin Harvey/NHPA
Page 20 © K. Ammann
Page 22 © Andrew Plumptre/Oxford Scientific Films
Page 23 © John Cancalosi/Bruce Coleman Collection
Page 25 © Noel B. Rowe
Page 26 © Konrad Wothe/Oxford Scientific Films
Page 29 © M. Watson/Ardea
Page 30 © Richard Packwood/Oxford Scientific Films
Page 32 © Martin Harvey/NHPA
Page 33 © K. Ammann
Page 34 © Christophe Ratier/NHPA

Page 35 © The Dian Fossey Gorilla Fund International
Page 37 © Ian Redmond
Page 38 © Andrew Plumptre/Oxford Scientific Films
Page 41 © Konrad Wothe/Oxford Scientific Films
Page 43 © Martin Harvey/NHPA
Page 44 © Andrew Plumptre/Oxford Scientific Films
Page 45 © Bruce Coleman Inc.
Page 47 © Steve Bloom/stevebloom.com
Page 48 © Michael Nichols/National Geographic
Page 49 © K. Ammann
Page 51 © Steve Bloom/stevebloom.com
Page 52 © Steve Bloom/stevebloom.com
Page 54 © Michael Nichols/National Geographic
Page 55 © Michael Nichols/National Geographic
Page 57 © K. & K. Ammann
Page 58 © Steve Bloom/stevebloom.com
Page 61 © Martin Harvey/NHPA
Page 62 © K. Ammann
Page 65 © Martin Harvey/NHPA
Page 66 © K. Ammann
Page 67 © Michael Nichols/National Geographic
Page 69 © Martin Harvey/NHPA

Printed in China

GORILLAS

Kelly J. Stewart

Colin Baxter Photography, Grantown-on-Spey, Scotland

Contents

Introduction 7

Origins 13

Characteristics 17

Life History 23

Ecology 31

Group Life 35

The Mating Game 45

The Future 59

Distribution Map 70

Gorilla Facts 71

Index 72

Introduction

I could tell the silverback was nearby because I suddenly smelled him – that pungent, sweaty odor of a nervous male gorilla. How close? Maybe 30 feet? I could see nothing except the dense green tangle of undergrowth through which I crawled.

I knew I was being too dogged. Nunki, as researchers called the silverback, was not yet at ease in the presence of observers, especially in poor visibility. He and his three females had been moving away from me for the last two hours, keeping just far enough ahead so that I couldn't see them. Now we were on a steep slope, and the gorillas were above me. Normally, I would have left, not wanting to push the animals or my luck. But I was keen to get a good look at one of the females, Petula, to see if she carried a new baby. The previous day, I thought I had glimpsed a tiny foot just under her arm, but I couldn't be sure. It seemed to me that Nunki had been unusually wary this morning, sometimes a sign that there has been a birth. Silverbacks seem especially protective of females with new infants. If only I could see Petula clearly. Maybe if I just crawled up a few more yards...

Nunki's roaring scream was the loudest I'd ever heard from a silverback. I remember thinking 'Maybe this is what they sound like when they aren't bluffing.' But it had been instilled in me to hold my ground in the face of a charge. Call their bluff and they will stop. They always had before. And I believe Nunki would have then, if it hadn't been so steep. He was coming too fast to stop. He landed on top of me, or rather, on top of the thick mat of vegetation that enmeshed me.

For what seemed an hour, but was probably less than a minute, life was suspended. The only sound was Nunki's hard breathing. My head was scrunched into the ground, but I could see the silvery-white hair of his back through the leaves, right there in front of my nose. Then he got up abruptly, moved uphill about 15 feet, and lay down and went to sleep. He had made his point.

The terrifying magnificence of a silverback's charge, his size and power, speak to something deep within us. Gorillas have grabbed humankind's imagination for centuries. It's as if they fulfill a need of ours to believe in a huge savage creature that is strange and wild, yet also humanlike, appealing perhaps to our inner ape. The Yeti, Bigfoot, King Kong, are all versions of the same theme. But the allure of gorillas does not lie only in their power and size. There is also the vulnerable, Gentle Giant factor. King Kong was a big misunderstood softie, after all. That morning when Nunki didn't stop, the worst thing I suffered was a pair of broken glasses. And by the way, Petula did have a new baby.

There are, of course, many reasons to study these apes other than their charisma. One of the most compelling is that, along with chimpanzees and orangutans, they are our closest living relatives. In anatomy, physiology, genetic make-up, they are more like us than any other creature on earth. If we can understand the forces that shaped the behavior of the apes, and how they differ from or resemble each other and us, then we might begin to understand where we came from, what our earliest ancestors were like.

A Short History of Research

The gorilla was the last of the apes to be 'discovered' by western science. Explorers and adventurers traveling to the forests of west-central Africa had been recounting tales of giant manlike creatures since the sixteenth century. But it was not until 1847 that these rumors were backed up by hard evidence. Thomas Savage, an American medical missionary, collected some skulls and bones of an unknown creature in Gabon. The published description of these specimens announced to the scientific world a new species of ape, the gorilla.

Over the next century, attempts to learn more about gorillas involved primarily killing them for museums or anatomical examination. While these collecting expeditions yielded valuable data on gorilla anatomy, morphology, and distribution, knowledge about their behavior and ecology lagged far behind. Most information concerning the habits of the animals was based on exaggerated hunters' tales, embellished with fanciful accounts from the local people. These stories overstated in particular the aggressive nature of the gorilla.

The French adventurer, Du Chaillu, was one of the worst culprits. In his lurid popular accounts in the 1860s of gorilla-hunting in west-central Africa, Du Chaillu refers to the adult male as a nightmare creature, 'an impossible piece of hideousness' that would not hesitate to kill a man.

Female and infant western gorilla.

It is likely that many silverbacks were gunned down in the midst of a roaring charge, protecting their group. Whether or not they were bluffing, we will never know. Hunters like Du Chaillu never gave the animals a chance to stop.

When people started observing gorillas rather than killing them, the animals' monstrous reputation slowly faded. It was not until the mid 1950s, however, that observational studies began to yield detailed information on the behavior and natural history of gorillas. The site for these studies was the forest-clad mountain range, the Virunga Volcanoes of Uganda, Rwanda and Democratic Republic of Congo (DRC). In 1902, Oscar von Beringe, a German officer on a military excursion, had 'discovered' gorillas here, at the most eastern edge, and altitudinal limit, of their range. The region became Africa's first national park, Albert National Park, in 1926, and its mountain gorillas, *G. beringei beringei*, became the best known in the world. The area is now divided among three contiguous national parks in Rwanda, Congo and Uganda. Most of the gorillas live in Rwanda's Parc National des Volcans and in Congo's Parc National de Virunga.

George Schaller's pioneering study in 1959 laid the groundwork for the most famous gorilla researcher, the American zoologist, Dian Fossey. In 1967 she pitched a tent at 10,000 feet (3048m) in the Virunga Volcanoes in Rwanda, thereby establishing the Karisoke Research Center. It was the start of more than 30 years of nearly continuous observation, one of the longest studies of any primate in the world. The descendants of those gorillas Fossey first contacted are still being observed today. Most of the information in this book on details of mating patterns and social behavior of wild gorillas comes from the 360 or so animals living in the Virunga Volcanoes.

After mountain gorillas, we know most about the other eastern subspecies, eastern lowland gorillas (*G. b. graueri*) in DRC or Congo as it is called. For over 30 years scientists have intermittently observed several gorilla groups in Kahuzi-Biega National Park, a region that includes both highland and lowland forest.

Far less is known about western gorillas (*G. gorilla*), also known as 'lowland' or 'western lowland' gorillas, largely because they are more difficult to study. Animals are harder to find and follow in lowland tropical habitat than in montane forest. In addition, gorillas in west-central Africa are traditionally hunted for their meat, making them more fearful and thus harder to observe. Over the last 15 years, however, the patience and persistence of researchers has paid off and we are learning more and more about gorillas all over Africa.

Unfortunately, there might be little time left to understand these magnificent relatives of ours. Gorillas face serious threats to their continued survival in the wild. Our species might manage to wipe them out, having only just got to know them; discovery in the 19th century, a glimmer of understanding in the 20th, and extinction in the 21st.

Origins, Taxonomy and Distribution

Gorillas Past and Present

Gorillas belong to the family of great apes, the Hominidae, that includes orangutans, chimpanzees and humans. The orangutans of Asia are classed in their own subfamily, Ponginae, while the African great apes, gorillas and chimpanzees, are classed together with humans as the Homininae. We are all Primates, the animal order that also includes prosimians such as lemurs and bushbabies, all monkeys, and the lesser apes, gibbons and siamangs.

Fossils tell us that ape-like creatures appeared in Africa more than 25 million years ago and spread throughout Asia and Europe. Paleontologists have identified over 15 genera of apes that lived at various times during the Miocene when places like Italy and Greece had tropical climates.

These creatures all went extinct and unfortunately, we have no fossils of the ancestors of gorillas or chimpanzees. Examination of the DNA of living apes, however, in combination with the fossil evidence from other taxa, give us some clues about their evolutionary history. Rough estimates suggest that the orangutan lineage diverged from that leading to the African apes and humans about 16 million years ago; the gorilla lineage separated from the chimp-human line about 10 million years ago, followed by the branch leading to chimpanzees about 8 million years ago. These dates are highly speculative and represent a compromise between disparate estimates. For example, the time for the divergence of the gorilla's lineage varies from 17 to 8 million years ago, depending on who you talk to. The picture is changing all the time with each new fossil and genetic analysis. Whatever the time periods involved, the order in which the separate lineages branched off from the 'trunk' leading to humans is now widely accepted. The startling conclusion is that gorillas and chimpanzees are more closely related to humans than either of them is to orangutans.

Today, gorillas, like chimpanzees, exist only in equatorial Africa. Unlike chimpanzees, which occur all the way across the continent from west to east, gorillas live in two widely separated regions, west-central and eastern Africa. Recent genetic work suggests that the gorillas in these two areas are different enough, at least in their DNA, to be classed as separate species. The western gorilla (*Gorilla gorilla*) is the most numerous and widespread, occurring in seven countries of west and central Africa (see distribution map on page 70). The small isolated population on the Nigeria/Cameroon border is considered by some people to be a separate subspecies (*G.g. diehli*) of western gorilla.

About 1000 gorilla-free kilometers to the east we find the eastern gorilla (*Gorilla beringei*) which is divided into two subspecies: the eastern lowland gorilla (*G.b. graueri*) in eastern Democratic Republic of Congo (DRC), and the rarer mountain gorilla, (*G. b. beringei*), in Rwanda, Uganda and eastern DRC. These two subspecies of eastern gorillas are completely separated from each other by unsuitable habitat including active volcanoes, lava forest, savanna, and human habitation.

It is only recently, with the advent of new genetic techniques, that scientists have considered eastern and western gorillas to be separate species rather than subspecies. Differences at the level of DNA, however, don't always translate into differences in anatomy, morphology or behavior. For example, the two species do not differ greatly in outward appearance, although experienced observers can tell them apart. Eastern gorillas are larger by about 45 lb (20 kg), and have longer, less flat-looking faces. The most obvious difference is their hair which, in mountain gorillas especially, is relatively long and jet black. By contrast, western gorillas tend to have shorter hair with more brown or reddish tints, especially on their heads.

Because eastern and western gorillas are comparable in many aspects of biology, I will use the generic term, gorillas, when conclusions about one species are applicable to the other. When differences occur, or when there is not enough information to justify generalizing, then I'll indicate the species.

A juvenile western gorilla. Some scientists now class eastern and western gorillas as two separate species, rather than subspecies. I have followed this new classification, but there is debate over the issue. The two have been separated geographically for many thousands of years. They differ in some traits but are similar in others.

Characteristics

Like all the apes, gorillas descended from an ancestor that lived in the trees and used its arms to hang and swing from branches, a mode of locomotion called 'brachiation'. This heritage is revealed in many of the gorilla's features such as the absence of a tail, relatively long arms, and shorter, rather bandy legs. Since their swinging ancestral days, gorillas have descended to the ground to become the most terrestrial of the apes. Although they often forage high up in trees and sometimes even sleep there, they spend much of their time roaming the forest floor.

A young western gorilla 'knuckle-walking'.

Gorillas, like their close cousins, chimpanzees, normally walk and run on all fours, supporting their forelimbs on their knuckles, rather than their palms. This unusual style of locomotion is called 'knuckle-walking'. On occasion, for instance when they are chest-beating, gorillas stand bipedally and walk or run for a few paces upright.

They are the largest of all primates. A wild male gorilla weighs, on average, 308 to 352 lb (140 to 160 kg), stands a little over 3ft 3in (1m) high when knuckle-walking, and over 5ft 6in (1.7m) from head to foot, when standing upright. Gorillas in zoos tend to be much

An adult male mountain gorilla has a longer face, and longer, blacker hair than his western counterpart whose crown is often russet.

heavier than those in the wild because of their richer diet and lack of exercise. You may have heard of gorillas weighing 500 lb (230 kg) and more. These weights come from captive animals who are way too fat!

Female gorillas are far smaller in stature and weigh about half as much as males. In fact, for many features of gorillas, the two sexes must be described separately because they are so different, a phenomenon known as sexual dimorphism. Most of their differences reflect a fact that is crucial to understanding these apes: males have evolved fighting ability. Apart from large body size, they have huge jaws and canine teeth, far bigger than females'. The jaw muscles attach to a distinctive crest of bone along the top and back of the male's skull, giving it the appearance of a Roman soldier's helmet.

The sexes differ, not just in weaponry, but also in features used in threat displays. After all, the ability to intimidate is a crucial aspect of competitive prowess. For example, while all gorillas beat their chests at times of excitement, the behavior is most developed in males who use chest-beating to announce their presence to their neighbors, or as part of an elaborate display during encounters with rival males from outside the group. The male's chest-beat has a hollow 'pok-pok' sound like a gigantic champagne cork going off. It is amplified by inflatable air pouches, which are extensions of the larynx or voice box. The distinctive sound travels impressive distances, unlike the slapping thumps made by chest-beating females and youngsters. The brawny chest of an adult male is completely bare of hair, a feature that probably improves its effectiveness as a drumming surface.

One of the most obvious differences between the sexes is the color of their backs. When a male attains full maturity, at around 15 years, the hair on his back becomes short, like a 'crew cut', and turns silvery white, hence the term 'silverback' for the fully adult males of both species. Females' backs do not undergo this transformation. Possibly, this silver

Eastern lowland male. Silverbacks do not climb trees
as often as smaller, lighter females and young gorillas.

saddle, striking against the darker surrounding hair, enhances the male's intimidating appearance by making him look bigger than he really is. It is a well-known optical illusion that breaking up a block of color appears to lengthen the object.

Why is fighting ability so important to males? Because they compete intensely with each other over females. The bigger and stronger the male, the more likely he is to attract and retain mates in a permanent group. Males who do not reside permanently with females have little chance of mating. Of course the male's power also serves as a defense against predators such as leopards or humans.

Gorillas don't show the 'whites of their eyes'.

In other characteristics, males and females do not differ. They have brown eyes and black leathery skin. Their eyesight and hearing is about the same as ours, but they attend to odor more than we do. For example, when they come upon an unfamiliar object, they will touch it and then sniff their fingers, and sometimes do the same with food items. It is not known whether their sense of smell is actually more developed than ours, or whether they simply pay more attention to it than we do.

Gorillas use a variety of vocalizations with which to communicate. The message of some of these calls is more or less obvious to human observers. For example, the mildly aggressive 'cough-grunt', the warning 'alarm bark', the lost infant's 'whimpering', or a 'fear scream' are given in specific situations and usually evoke specific responses. Far more mysterious are the quiet, belch-like grunts that gorillas give throughout the day in different contexts. Animals might grunt while they are feeding, moving, or just lying down and resting. The only obvious response to these sounds is the occasional answering grunt. These vocalizations might convey a very general message, such as 'I am here' or 'I am about to change activity', in which case they could function, in part, to coordinate group movement and activity. For instance, towards the end of a mid-day rest period, gorillas start grunting more frequently, as if to indicate that they are ready to end the siesta and move on. Even when the animals are doing nothing but lying there, an observer can often tell when the rest period is about to end, just from the increase in 'conversation'.

The vocalizations of apes appear to have little in common with human speech. Scientists working with captive animals, however, have investigated apes' mental capacity for language by teaching them to communicate with humans using hundreds of symbols and hand signs. While most of this research is conducted on chimpanzees, who are more common in captivity than are gorillas, the one gorilla involved in similar work has also learned to use sign language. Although scientists largely conclude that this is not equivalent to human language, the work has opened a window into the minds of apes and suggests a certain degree of complex cognitive abilities. In general, it appears that the great apes are more intelligent than monkeys, even though their brains, relative to body size, are not notably larger than those of monkeys.

In the wild, gorillas appear less mentally advanced than chimpanzees, due to their apparent lack of tool use. Chimpanzees, in contrast, frequently use tools to obtain food. They prepare twigs, for example, to stick into termite mounds and 'fish' for the insects, and use rocks to crack open nuts. It is not clear whether this difference between the apes reflects primarily contrasting levels of brain power, or simply differences in diet or strength. Gorillas might not need tools because they rarely eat hard-to-get-at foods and when they do, brute strength suffices. Chimpanzees 'fish' for termites. Gorillas smash apart the mound.

Life History

Growth

It takes many years for the striking sex differences in adult gorillas to develop. At the beginning of life, males and females look very similar. Newborn gorillas are tiny compared to their mothers, weighing in at about 4lb 8oz (2.1 kg). This is about half as much as the average human infant weight at birth, even though adult gorilla females are quite a bit heavier than the average woman.

During the first four to six months of life, the infant gorillas are in continuous contact with their mother's body, clinging to the hair of her belly and chest, or being cradled in one arm. They then begin making tentative, wobbly ventures away from their mother, as they start to explore the physical and social

A young western infant snoozes on its mother's back.

world surrounding them. By the end of their first year, they travel on their mother's back like little jockeys, and spend much of their time in wild games of wrestling and chasing with other youngsters. By now they have learned to eat the same food their mother eats, although they are not yet adept at preparing some of the more difficult items, such as stinging nettles or prickly thistles. These techniques are learned during their second year, by

A mountain gorilla cradles her newborn infant. Its pinkish-gray skin will gradually turn black.

which time they are traveling independently more often than riding on their mother.

Gorilla infants are nursed regularly during their first two years. Although they can probably feed themselves by 24 months, they are not fully weaned until two-and-a-half to three years old. Even then, they continue to share their mother's nest at night, until the next infant comes along.

Males and females closely resemble each other during their juvenile period (three to > six years) and adolescence (six to < eight years). In fact, in young mountain gorillas, whose long hair hides telltale genitals, the sexes are rather hard to tell apart. During my study at Karisoke Research Center, I was shocked more than once when one of my study subjects 'changed sex'. In the most embarrassing case, an adolescent 'male' gave birth!

Males begin a growth spurt at around six years, but don't appear obviously different from females until about eight or nine when they move into the 'blackback' stage. This is equivalent to the teen years, during which they gradually develop the distinctive features of adult males. The process is a slow one, and it isn't until 13 years or so that the hair on a male's back begins silvering, and not until 15 that he is fully mature.

Reproduction

Although blackbacked males occasionally mate with females, we do not know at what age they start producing viable sperm. In captivity, there are reports of fertile matings by males between six and nine years old. In the wild, in both species, it is unlikely that males would sire any offspring before full adulthood, because they would be unable to attract fertile females.

Females reach puberty and show an interest in sex at around six to seven, but do not

A juvenile mountain gorilla from Rwanda, where researchers at
Karisoke Research Center have studied wild gorillas for over 30 years.
Compare this animal to the less shaggy western juvenile on page 15.

*Lactating mothers, like these two western gorillas, can often
be identified by their enlarged breasts and long nipples.*

give birth for the first time until they are 10. This interval between puberty and fertility is known as 'adolescent sterility', and has also been reported for maturing chimpanzee and human females. It may have evolved as a means for subadult females to learn about the other sex and choose an appropriate mate before taking on the load of motherhood.

For fully adult female gorillas, sex is a relatively uncommon activity. Unlike human females, who mate throughout their menstrual cycle, gorillas tend to mate only when they might conceive, which is during ovulation. As in humans, this occurs once a month for only two to three days. Then, females typically initiate mating by approaching males in a coy, hesitant manner, giving characteristic vocalizations.

Females usually conceive after just two or three sexual cycles, and during their pregnancy which lasts for 8.5 months, they sometimes mate, but only infrequently and sporadically.

I have witnessed two births in the wild and both were low-key, undramatic affairs. The females both delivered in the afternoon in close proximity to other group members. In each case, detectable labor lasted less than 30 minutes, during which time the female behaved restlessly as if she couldn't get comfortable, frequently changing her seated position, and reaching down to touch her perineum. In the first case, I didn't fully realize what was happening until the infant was born! One of the mothers delivered the placenta three minutes post-partum, and ate the entire thing with great gusto.

After giving birth, a female's sex life is put 'on hold' for the next two and a half to three years. The action of suckling her infant maintains lactation and suspends ovulation and sexual receptivity. In mountain gorillas, this results in an average birth spacing of four years, ensuring that a female does not produce another infant before her current one is independent. In a well-fed, healthy population like that of the Virunga Volcanoes, a female is considered reproductively successful if she has four offspring that survive to adulthood. Many don't get past infancy.

Mortality

As for many other wild animals, infancy is the most dangerous time of life for gorillas, especially early infancy. In mountain gorillas, nearly two out of every five infants (40 per cent) die, most during their first year. Unrelated silverbacks kill some of these infants, but for most deaths, we don't know the cause. One day the baby is in its mother's arms, and the next day it's gone.

Once a gorilla has made it past infancy, what might kill it? Some animals die of disease. Autopsies on young and old mountain gorillas reveal a variety of illnesses including cancer, gastroenteritis, and, most commonly, pneumonia. In west-central Africa, gorillas have died from Ebola. The disease can also devastate chimpanzee populations, and is the same form that kills humans. In fact, recent Ebola outbreaks in Africa among humans can be traced to the eating of diseased ape flesh.

Silverbacks can die from wounds received in fights with other males, but this is extremely rare. As for predators, gorillas are too big for most animals. Leopards have been known to kill them, but there are only a handful of documented cases. Interestingly, adult males are among the victims. One can only imagine the fierce struggle that accompanied these encounters. It must be a very hungry leopard, indeed, that takes on a silverback. Females and youngsters, however, would undoubtedly be more vulnerable to leopards were it not for the silverback's protection. By far the most serious predator of gorillas is the human hunter.

It is difficult to be precise about gorillas' lifespan. Wild gorillas that make it to adulthood probably live to their mid-20s to early 30s. Adults older than 35 are considered elderly and those of 40, unusually old. The oldest gorilla in captivity was a male who died at 54 in Philadelphia Zoo.

In the wild, males generally have a shorter life span, by a few years, than females.

Ecology

Salad Days

People studying wild animals know the truth of the saying, 'you are what you eat'. An animal's diet has a huge impact on other aspects of its life, including home range size, daily traveling distance, and degree of sociality.

The most important food fact about gorillas is that they are able to subsist on foliage, that is, the leaves, stems, and roots of vines, herbs, trees, and grasses. The drawback of such a diet is that it's fairly low quality compared to, say, ripe fruit. A gorilla must consume an enormous amount of greenery in order to obtain the necessary energy and nutrients. The advantage of living on foliage is that there's lots of it, available year-round.

Mountain gorillas are the salad-eaters supreme. In the Virunga Volcanoes, the region from which they are best known, there is virtually no fruit because of the high altitude. Most of the forest lies above 8000 ft (2440 m) with the tallest peak topping 14, 500 ft (4507 m). The gorillas here spend much of their time at or below 10,000 ft (3030 m), where the slopes are covered in dense stands of bamboo, or in a thick tall carpet of secondary vegetation – nettles, thistles, gallium, giant celery. The animals are literally surrounded by their food, an abundance that affects the amount of land they need.

In both east and west-central Africa, a gorilla group frequents a particular region of forest known as its 'home range'. Although the general size and location of a group's range can be stable for months and even years, these are not territories in the classical sense. Gorillas do not defend boundaries against intruders, and group ranges overlap extensively with those of their neighbors.

Mountain gorillas have the smallest home range of any gorilla subspecies, about

Mountain gorillas are highly folivorous and surrounded by food.

3 sq miles (8 sq km), and travel only 550 yd (500 m) on an average day. This is partly because they don't have to search far to find food, but also because they need time-out from foraging to rest and digest their leafy fare. But the Virunga Volcanoes are an extreme habitat whose denizens have no choice but folivory. Other regions offer more varied menus, and animals' diets differ accordingly.

In striking contrast to the Virunga habitat are the fruit-rich lowland forests of west and central Africa. Here, western gorillas eat a wide array of fruits that, in some areas, make up over 50 per cent of their diet at certain times of year. This food is not, however, ubiquitous like ground-growing foliage. Fruit trees are dotted here and there throughout the forest, and the amount of food they contain can be unpredictable. Ripe fruit is harder to find than leaves and herbs, but it is worth the extra mile for the sugars and other nutrients it

Mountain gorillas spend almost half the day feeding.

contains. Consequently, western gorillas have larger home ranges than do mountain gorillas, and travel almost three times as far each day.

Fruit-eating also influences group cohesion and sociality. In western gorillas, for example, group members are far more spread out during feeding than in mountain gorillas. In some regions, groups have been known to split into subgroups and then rejoin, often in response to fruit availability. Despite these frugivorous tendencies, however, folivory is still a crucial fact of life for western gorillas.

When the going gets tough, in the non-fruiting season or when fruit crops fail, western gorillas switch over to lower quality, but more abundant foliage. This is why they prefer relatively open forest like the type that grows on ridge-tops, or hillsides, where there are gaps in the tree canopy, and a dense layer of undergrowth. Gorillas are less numerous in dark looming primary forest with its bare floor.

Western gorillas frequent another type of habitat for the vegetation it offers. In some regions of west-central Africa, the forests are interrupted by large clearings of permanent swamps that attract remarkable numbers of large mammals, including gorillas. They sometimes spend all day in water up to their elbows, in the company of elephants, buffalo and giant forest hogs, eating the sedges, grasses and herbs that grow in the swamp.

Wild gorillas are almost complete vegetarians, apart from the occasional meal of ants or termites. In comparison, their close cousins, chimpanzees, are well known for their monkey-hunting, monkey-eating habits.

Gorillas' folivory enables them to live in more or less permanent groups, since there will usually be enough food at any one spot for all to get their fill.

In eastern Congo, gorillas may feed in grassland.

This contrasts with the social orders of the other great apes, who rely more heavily on ripe fruit. Chimpanzees have a fission-fusion social system where the size and membership of associations can change over days or even hours, depending on what they are eating. A fruiting tree might attract a large party of chimps who then split up to forage alone when the fruit is finished. Orangutans are the least social ape, leading essentially solitary lives.

Group Life

Group Size and Membership

Gorillas live in cohesive groups whose membership, apart from changes due to births and deaths, can remain constant for months and even years. The typical gorilla group contains one fully adult male, two or more adult females, and their offspring of various ages. Groups sometimes have more than one silverback, although the frequency of this varies with region. In mountain gorillas, 26 to 40 per cent of groups are multimale, whereas in eastern lowland and western gorillas, the proportion is closer to 10 per cent.

Adults sometimes play with younger animals. This silverback shows an open-mouthed 'play face'.

Across Africa, gorilla group size averages about 10 animals, but can vary from two – for example a newly formed group of one male and one female – to an astonishing 49. This record is currently held by one of the Karisoke study groups in the Virunga Volcanoes.

A gorilla group's size and structure results from births and deaths as well as the immigration and emigration of individuals. At maturity or later, both males and females tend to leave the group they were born in. Females immediately join another group or solitary silverback, while males usually travel on their own and do not breed unless they manage to acquire females from other groups. Fully adult males cannot join breeding groups because of resistance from the resident silverback.

Thus, while all adult females reside in groups, only some adult males do so. A system like this, in which some males have several mates, while others have none, is called 'polygyny' and is quite common among social mammals. It is often associated with competition among males for females, and with sexual dimorphism like we see in gorillas.

Social Relationships Within a Group

Nearly all our information about group life in the wild comes from the mountain gorillas in the Virunga Volcanoes. For over 30 years, observers based at the Karisoke Research Center have been following the fortunes of several habituated groups.

Habituation is a process whereby observers accustom 'naïve' gorillas to their presence. It takes patience and perseverance on the part of the observer, locating the same group day after day, behaving in a low-keyed, non-threatening manner, and attempting to stay with the animals as they move through the forest. Initially, the silverback charges or screams and the group flees. Next comes the curiosity stage, when the animals linger at a safe distance, beating their chests in excitement, and peering through the foliage to glimpse their persistent intruder. By now, the silverback has scaled down his warnings to 'alarm barks'. Eventually the animals lose interest and simply get used to the presence of observers, paying them little attention. The process can take many months or years and sometimes fails completely. Its success depends partly on the nature of the silverback. If he is a calm animal who habituates quickly, then so will his group. Some silverbacks never do accept observers. In the Virunga Volcanoes, there was a male named Brutus, whose terrifying charges were legendary. For over 10 years, researchers tried to habituate Brutus and his group, but to no avail.

When I studied mountain gorillas, I was interested in the nature of social relationships within a group. Each morning, I would set off from camp, and walk through the forest to where the animals had last been seen. From there, I would follow their winding trail of flattened vegetation until I reached the group, and would then remain with them for hours, recording who did what to whom, how often.

Most days followed a similar pattern. A gorilla group rises with the dawn and spends

In some regions of west-central Africa, gorillas frequent large swampy clearings in the forest (also see photo on page 34). Several groups may be present at the same time, often ignoring each other to feed on roots, sedges and herbs.

*Silverbacks, like this mountain gorilla in Rwanda, are tolerant and protective
of the infants in their group, who are likely to be their offspring. In mountain gorillas,
infants that lose their mothers become increasingly attached to the dominant male.*

much of the morning moving through its dense green world, feeding. Animals are spread out when they feed, and very often, an observer cannot see more than one individual at a time.

At midday, the group normally has a rest period of about an hour, when animals come together and play, groom each other, or just doze. The afternoon is taken up with more travel and feeding until dusk, around 6:00 p.m., when the group stops and beds down for the night. Each gorilla, except infants who sleep with their mothers, makes a rough nest by bending branches and leaves underneath them. Nest building is a common feature of all the great apes, but while chimps and orangutans always make their nests high up the trees, gorilla nests tend to be closer to or on the ground, especially those of adult males.

If you watch a gorilla group for even a short time, it becomes clear that the silverback is the focus of other animals' attention and activity. Group members gravitate towards him during rest periods, and follow him when he moves off, treating him as a leader. Adult females often spend more time near him than near any other adult, and have more friendly interactions such as grooming. It is evident from observations like these that the affinity between adult females and the silverback is the social backbone of a group. But it is certainly not the only tie that binds.

As with many primates, the bond between mother and offspring is a strong one, lasting well beyond the weaning stage. This is clear, not just from friendly interactions, but from the way that mothers support their immature offspring during their aggressive encounters.

Besides their mother, the closest adult relationship for most immature gorillas is with the leading silverback, to whom an attraction develops early in life. It is not unusual to see an adult male surrounded by a cluster of infants and juveniles, with their mothers nowhere in sight, as if the silverback were a sort of day care center.

Most dramatically, if a female dies when her offspring is still unweaned or newly weaned, the 'orphan' turns, not to other females as surrogate mothers, but to the silverback, becoming his little shadow during the day and sharing his nest at night. Silverbacks appear to be extra tolerant and protective of motherless infants in their group. The youngest animal to survive its mother's death that we know of was less than two years old, well before the usual age of full weaning.

Dominance rank in mountain gorillas is generally based on age, up until adulthood. Then males, because of their larger size, are able to outrank older females. The silverback is the most dominant animal in the group. He can supplant others from a feeding spot by merely approaching them, or he can quickly break up fights with a short, sharp vocalization. He almost never receives threats in return, but instead is deferred to by subordinates with the grumbling, humming vocalizations that indicate appeasement.

Life gets more complicated when there is more than one silverback, a situation that arises when males born in the group remain there as adults. In this case, there is a dominance hierarchy among the males, normally based on age. Relations between them can be tense, especially when females are in estrus. Even then, disputes rarely escalate into actual wounding fights. Despite their differences, males in the same group will jointly defend females and young against outside threats such as other silverbacks or predators. The key to this tolerance is the fact that they are related to each other.

In mountain gorillas, a silverback's tenure as alpha male usually lasts about 10 years. Eventually, as he ages, his top position is usurped by the next youngest male in line, very often his son. In one father-son pair, this process took about a year. An older silverback that drops in rank is not ousted by the younger male, but remains an integral member of the group.

Among the adult females, dominance relations are not distinct, partly because the silverback's intervention in their disputes tends to even out their competitive differences. For example, a young, small female might be able to hold her own against an older, larger animal with a little helpful intercession from the male. It is possible that a female's relationship with the dominant silverback might influence her social standing. In fact, one of the things that adult females fight about is access to the male. Who gets to sit next to him today?

In the Virunga Volcanoes, the gorillas' habitat is completely surrounded by farmland, often terraced to accommodate the steep slope. Mountain gorillas sometimes walk along the edge of cultivated fields, but stick close to the forest. In Rwanda, gorillas do not raid people's crops, which consist mainly of potatoes and beans. In other regions of Africa, however, gorillas occasionally destroy cultivated banana trees to eat the pith.

Most of the time, adult females tend to ignore one another, unless they are close relatives, such as mother and daughter or sisters. Then, close bonds are evident. But because females habitually transfer between groups, it is not the norm to find adult relatives living together.

Group Fission

When births exceed deaths and immigration of females exceeds emigration, a gorilla group can become very large. In this case a split might occur as long as there is more than one mature male. When a group fissions, one of the silverbacks 'buds off' with a portion of the adult females and their offspring, while the rest of the females stay with the other adult male. The process has been documented in both mountain and eastern lowland gorillas.

In one fission of a Karisoke study group, the females tended to choose the male with whom they had been friendliest before the split.

Other Regions of Africa

This general description of group life from the Virunga Volcanoes applies in large part to eastern lowland gorillas (G.b. graueri).

For western gorillas, unfortunately, we do not have anywhere near this amount or detail of information. Basic features, such as group size and composition and dispersal patterns, appear to be similar to those found in eastern gorillas. However, beyond this there is little we can say for certain, especially when it comes to social relationships within a group. Several researchers have reported more flexible grouping tendencies in western gorillas, with groups splitting into subgroups, sometimes for days at a time, and then reforming. This may be related to the abundance of fruit, but might be restricted to groups with more than one silverback. Only continued studies in west-central Africa will fill in this gap in our understanding.

The Mating Game

The movements of individuals into and out of groups are driven largely by the search for mates, a quest that begins in young adulthood. When the sons and daughters of a gorilla group reach maturity, they can follow very different paths in pursuit of a breeding career. The following account outlines what we know from observing individual male and female mountain gorillas throughout the sometimes soap-operatic course of their lives.

A mountain gorilla silverback.

Females

Young adult females either stay in the group in which they were born, known as their 'natal group', or they may emigrate to breed elsewhere. Among mountain gorillas, about 50 per cent of daughters leave before they have reproduced, another 30 per cent have at least their first infant at home, and then leave, and only about 20 per cent remain to spend their breeding lives in their natal group.

We do not know what determines if or when a daughter will leave her natal group, but avoiding close inbreeding, or 'incest', seems to be one factor. It is possible that when a female reaches adulthood, her only choice for a mate will be her father. This is because the average tenure of 10 years for an alpha male is longer than the time it takes daughters to mature.

Males

Like females, young adult males either remain at home to breed, or disperse. Major influences on a male's 'decision' to stay or leave are the number and age of other adult males in the group, as well as the number of adult females. These factors will determine a young silverback's breeding opportunities.

As males mature from blackbacks into silverbacks, their sex life is limited by aggression from the dominant male. Subordinates sometimes sneak matings, literally behind the bushes, but these trysts are frequently interrupted by the older male, especially when they involve fully adult females.

Aggressive displays include impressive runs.

Despite this hindrance to breeding, a young silverback could benefit by biding his time in his natal group, making do with surreptitious matings. If the dominant male, very often his father, is getting on in years, and if the younger male is next in line, it might not be too long before he can reverse rank with his elder and advance to alpha position. Even if there are other up and coming males in the group, the alpha son may be able to monopolize most of the mating, at least for a few years.

We do not know how mating is divided among the silverbacks of a multimale group, but it is not determined just by them. Females have a say in the matter. Do some females mate with all males, while others favor just one? What male traits besides dominance rank guide a female's choice? How strongly do the aversions of close kinship affect both male and female mate preferences? The relatively new field of DNA analyses can help solve some of the puzzle by revealing the paternity of animals born into multimale groups.

For some young males, waiting around for breeding opportunities simply doesn't pay off. If the dominant silverback is in his prime, or if a male has older half brothers or cousins ahead of him in the dominance 'queue', then his mating chances will be few. In such a situation he may opt for greener pastures and leave.

When males emigrate, they do so voluntarily, becoming more and more peripheral to their group until they are alone. During this solitary stage, which sometimes lasts for years, males make long forays way outside their normal range. The point of these 'safaris' is to find females through close encounters with groups.

Lone males travel fast and far, sometimes even during the night, to pursue this goal.

Encounters between a group-living silverback and an outside male can be times of high drama and often violence. Both males perform ritualized threat displays that include hooting, chest-beating, smashing vegetation, charging and stiff-legged strutting. It's a show of bravado, demonstrating health and strength,

Confrontation in a strut, but with gaze averted.

aimed at both intimidating the rival male and attracting females. Even encounters between two groups can be dramatic as both silverbacks try to lure away the other male's mates while retaining their own.

Most encounters do not go beyond threat and the groups part with no harm suffered. Occasionally however, displays can turn into fierce fights in which males are seriously wounded and sometimes killed. Dian Fossey once found a silverback's skull with the canine tooth of another adult male imbedded in the bone just above the eye socket.

There is much at stake, especially for a lone male. His only chance of breeding is to attract females away from other silverbacks. This is how new gorilla groups originate. Joining a pre-existing group is not an option because of resistance from the resident silverback.

Going it alone is a risky strategy for a young male. For example, one individual became a solitary male when he was 14 and spent the next three years alone, vigorously seeking out groups and challenging their silverbacks. He never obtained more than two females, and spent only three years as an 'attached' male. When he died at age 20 from wounds incurred in a fight with another silverback, he left behind no surviving offspring.

Like any risky business, however, the lone-male option can have huge payoffs. In 1974, Nunki appeared suddenly in the study area as a lone silverback, and immediately acquired two females. In six years he had increased his harem to six females, and fathered 14 infants, eight of which survived into adulthood.

Infanticide – the Dark Side of the Gentle Giant

A major benefit for a male like Nunki is that he is the sole silverback in the group, and therefore the sole breeder. There is a drawback, however, to being a single adult male with a harem. This was obvious when Nunki died of illness. The group was left without a silverback, and disintegrated, the four adult females scattering to join different groups or lone males. Three of these females had infants, all sired by Nunki, who were killed by the new silverbacks.

The killing of unrelated infants is a male mating strategy that has been documented in many other animal species, including other primates. In the Virunga Volcanoes, infanticide accounts for almost 40 per cent of infant deaths. It has evolved because the males who kill other silverbacks' infants are more successful breeders than those who don't. A female that loses the infant she is still nursing will stop lactating, and as a consequence, return to sexual receptivity within a short time. Then, she will mate with the new silverback. If the male did not kill her infant, he would have to wait two or more years before she was ready to mate.

The fact of infanticide is difficult to reconcile with the image of a silverback reclining during a rest period, while his group's infants clamber about on his body in play. It is thought

that a silverback judges whether or not an infant could be his by his familiarity with the mother, especially whether or not he has mated with her, and possibly how often and when.

One might ask why Nunki's females with young infants would join a strange silverback only to have their infants killed. Why not just live without a male until they wean their offspring? The answer is that a female in this situation is cutting her losses. An unfamiliar silverback, a solitary male for instance, would eventually find her and try and recruit her as a mate, killing her infant in the process.

In mountain gorillas, infanticide is a relatively rare event because females with young infants stay put in their groups with their leading males. The most common situation in which infanticide occurs is when a leading silverback dies and there is no younger male to take over from him.

By contrast to the fallout from Nunki's death, when a silverback in a multimale group dies, there is usually a smooth transition to the next male in line. He is familiar with the adult females and has probably mated with some of them. The infants that aren't his own are relatives such as siblings, nephews and nieces. It would seem that breeding in a multimale group would be best for females.

Why, then, don't all females transfer to multimale groups? Why do over half of gorilla groups have only one adult male? One answer is that multimale groups take so long to develop. Even if a silverback's first infant with his first female is a son and it lives, it will still be 15 years before he can effectively defend a group of females and offspring.

A Band of Bachelors

When Nunki died and his group disintegrated, his juvenile and adolescent sons and daughters followed their mothers into new groups. While the immature females were accepted by the new silverbacks, eventually becoming their mates, the immature males were driven out, obviously viewed as future competitors. Young males like these without a group, sometimes band together, joining up with other drifters. One such bachelor group in the Virunga Volcanoes stayed together for seven years, with males dispersing to become solitaries as they reached full maturity.

Other Regions of Africa

It takes many years to collect the reproductive histories of individuals, and we know less about the mating strategies of gorillas from other regions of Africa. From the information available, there appear to be general similarities with some differences.

We have much to learn about western gorillas.

Male and female dispersal from groups, female transfer, and solitary males occur in eastern lowland and western gorillas. By contrast, only in mountain gorillas has infanticide been reported. In western gorillas, the difference could be due to the paucity of long-term observations on known animals. For the other eastern subspecies, however, lack of data is not a satisfactory explanation. For almost three decades, scientists have been observing five habituated groups of eastern lowland gorillas in Kahuzi-Biega National Park in DRC, or Congo as it is usually called. Although the work has suffered frequent interruptions due to the chronic political instability of that region, nevertheless, researchers have been able to follow known groups and individuals over many years.

The lack of observed infanticide in these gorillas is especially puzzling, since scientists

Poachers often capture young gorillas after killing their parents, and try to sell them as pets. A handful of projects in west-central Africa rescue orphaned apes, like this juvenile in Gabon, rearing them in naturalistic enclosures with the hope of one day returning them to the wild.

have witnessed numerous occasions in which it was expected. For example, females have transferred from one group to another with young unweaned infants who were not harmed by the new silverback that the mother joined.

A possible explanation for some of these cases is that the transferring female and her 'new' male had previously been in the same group. Thus, the new silverback was actually familiar with the female and maybe even related to her infant, a situation that would inhibit infanticidal tendencies. Another possibility is that, for some reason, there are relatively few unmated males (the most likely infant-killers) in the gorilla population of Kahuzi-Biega, and therefore relatively low levels of male-male competition for mates. This possibility may relate to another unexpected observation from this region.

In striking contrast to mountain gorillas, when the silverback of a one-male group dies in Kahuzi-Biega, his females do not always join other males, as happened after Nunki's demise. Instead, females and their offspring can remain together as a cohesive 'widow' group for months and in one case, over two years. This group, which by then had no unweaned infants, was finally joined by a solitary male who became the leading silverback. It is hard to imagine a group of mountain gorillas remaining 'unattached' for so long.

How did these gorilla females and their immature offspring behave without a fully mature male in residence? Apparently quite normally, although there was an interesting change in their sleeping behavior at night. They went from building their nests mostly on the ground, to building them mostly in the trees, as if they felt more vulnerable to predation without a silverback in attendance.

Understanding the difference between gorillas from different regions of Africa is a major goal of present and future studies. Unfortunately, it may be many years before we can solve the puzzles presented by research on eastern lowland gorillas. Since 1999, the war raging in Congo has taken a terrible toll on wildlife, and four of the five habituated gorilla groups in Kahuzi-Biega National Park have been wiped out, slaughtered by poachers or soldiers.

The Future of Gorillas

In 1979, one of Dian Fossey's habituated study groups was attacked by poachers. They killed both silverbacks, cutting off their hands and heads for sale as trophies. They killed an adult female and her young juvenile, while attempting to capture the youngster for sale as a pet. The remaining females of the group dispersed, two of them losing their young infants to infanticide.

Tragic though this incident was, it marked a turning point in the fortunes of the mountain gorilla. At the time, the animals were in dire straits. Although their habitat was enclosed in national parks in Rwanda and Congo, protection was minimal. The gorillas faced habitat loss as well as poaching, and their population had been declining steadily to half the numbers of 20 years before.

The Mountain Gorilla Project

The violent demise of Fossey's study group generated a massive publicity campaign, resulting in an internationally funded conservation effort that helped save the mountain gorilla. The Mountain Gorilla Project, based in Rwanda, included three prongs of operation: improving park protection, initiating a conservation education program, and developing gorilla-based tourism. This last venture entailed habituating several groups of gorillas to small parties of tourists who could observe the animals at close range. The aim was to make gorillas and their habitat economically important to the country.

No one could have predicted how spectacularly successful it would be. Not only did tourism bring desperately needed revenue to the National Park, but in a few years, the gorillas became the third-largest earner of foreign currency in Rwanda. Suddenly, the country could not afford to lose them.

With the tangible economic benefits of tourism combined with improved park protection and conservation education, the MGP helped to turn around the fortunes of the mountain gorilla. Periodic censuses show that gorilla numbers in the Virunga Volcanoes have increased from about 270 in the late 1970s to almost 360 in 2000. The only other population of mountain gorillas are the 240 or so animals in the Bwindi Impenetrable Forest in Uganda, where conservation measures similar to those in the Virunga Volcanoes have been effective.

Troubles with Tourism

I must make a slight diversion here to consider the pros and cons of tourism. While the mountain gorilla tourism program has become an icon of successful conservation, it has its drawbacks. There are risks inherent in any situation in which groups of strangers come into daily close proximity with wild animals. These include disturbing the animals, or exposing them to foreign pathogens to which they have no resistance. Gorillas are susceptible to most of the same diseases that we get. In addition, habituation itself might make gorillas more vulnerable to poachers by lowering their vigilance or wariness. While it is crucial to recognize these risks and do everything possible to minimize them, we must always weigh them against the benefits of a tourism program. For example, balanced against the potential danger of lowered vigilance is the fact that guards and guides from the National Park constantly monitor a gorilla group that has been habituated for tourism. This is effectively a daily anti-poacher patrol in the gorillas' neighborhood, probably the most effective immediate form of protecting the animals.

As for the risks of disturbance and disease, these can be reduced through a well-regulated visitation program. The present guidelines, originally established in Rwanda, include restrictions on the number of visits per day, the size of the tourist party, the timing of the visit, and the distance tourists must keep between themselves and the animals. In addition, anyone with a cough or cold is not allowed to go to the gorillas. Veterinary

Tourists to the Virunga Volcanoes are virtually guaranteed close views of habituated mountain gorillas. A similar program in Uganda has been just as successful. Visitors come away awe-struck and deeply moved by the experience. Many potential risks of tourism can be minimized through strict control of visitor numbers and behavior.

projects in Rwanda and Uganda monitor, with non-invasive techniques, the health of the habituated gorilla groups.

Unfortunately, there is always pressure to relax the rules in order to increase tourist volume. By and large this pressure has been resisted. To date, we have no evidence that tourism in Rwanda disrupts the gorillas' lives. For example, their daily activity patterns, ranging and feeding behavior and reproductive rates do not differ significantly from those of the Karisoke study groups. Nor has there been a clear case of animals contracting a disease specifically from tourists.

There are other, more general drawbacks to tourism. The fickleness of the industry, for instance, is well known. When political unrest or war erupts, even in a neighboring country, the first thing to disappear are tourists who cancel their reservations and travel elsewhere. This can be devastating if all conservation effort and its justification depend on tourist revenue. Conservation programs should never put all their eggs in one tourist basket. The Mountain Gorilla Project would not have been the success it was without the improved park protection, publicity campaigns, and conservation education programs that ran in parallel with the tourism initiative.

In summary, there are, and always will be, drawbacks to gorilla-based tourism. The risks to the animals are greatest for mountain gorillas who are easy to find and follow because of the nature of the habitat through which they move. In most regions of west-central Africa, the gorillas travel too far or are too difficult to find to allow the sort of close-up, camera-friendly visits that are virtually guaranteed a tourist who visits a mountain gorilla area. Nevertheless, almost all conservation projects focused on western gorillas include development of some level of eco-tourism, with a view of gorillas as the peak experience.

Despite the risks, when run properly, eco-tourism is one of the least destructive forms of nature exploitation that we know of. It works as conservation because the people of the country benefit economically if the habitat and its wildlife are left intact rather than destroyed. Gorilla-based tourism is here to stay, and conservationists must work to minimize its risks. It's not ideal, but it's better than hunting or logging.

Current Threats

The success of the Mountain Gorilla Project does not, unfortunately, ensure the animals' future survival. After all, the two separate populations of mountain gorillas together comprise only about 600 animals. Furthermore, in 2002, there was a resurgence of poaching in the Virunga Volcanoes, apparently to capture infants for sale to pet owners or unscrupulous foreign zoos.

As for gorillas elsewhere in Africa, their survival is no less tenuous. The IUCN Red List classifies gorillas as endangered in all parts of their range. The problems they face are only going to worsen with time, especially since only 10 per cent of western gorillas exist within officially protected areas like reserves or National Parks.

As for many species, the most serious long-term threat to gorillas is loss of habitat. Across Africa, there is a direct relation between human population density and rates of forest clearance, much of it for agriculture. People need land to feed themselves. On top of this, commercial logging exerts an increasingly heavy and negative impact on gorillas' forest habitat.

Hunting poses the most serious immediate threat. Except for mountain gorillas, the animals are killed for their meat everywhere they live. In 1980 I was in a restaurant in Gabon where the head of a silverback was mounted on the wall, and the rest of the animal was on the menu. Hunting for food is an important tradition in Africa, but in many areas it is no longer sustainable because of the increase in human population and the proliferation of guns. To worsen matters, commercial logging roads open up large tracts of previously inaccessible forest to hunters, and logging companies themselves feed their crews off the forest. The result has been the commercialization of the 'bushmeat' trade over the last 10

years. It is big business. Trucks thunder into the forest, load up with dead animals of all kinds, and transport the meat to urban centers. Marketplaces are full of gorilla and chimpanzee body parts, and it is thought that hunting could well be wiping out local ape populations.

The political instability and war that plague many regions of Africa have also taken their toll in recent years. The tragic long war in Democratic Republic of Congo (DRC), for example, has had a terrible impact on eastern lowland gorillas. Their refuge, Kahuzi-Biega National Park, now contains thousands of people in makeshift settlements, living off the meat of the forest, or digging for coltan, a mineral used in the making of cellphones. The elephants have been all but wiped out and over half of the gorillas have been killed. The poverty and starvation that accompany war force local people to exploit whatever resources they can. Anarchy and easily obtained guns increase the problem.

Gorilla meat for sale in a market in west-central Africa.

In 1996, published estimates of gorilla numbers across Africa indicated about 111,000 western gorillas, most of them in Gabon and Republic of Congo, and around 10,000 eastern lowland gorillas in DRC. Unfortunately, these widely quoted numbers underestimated the impact of ever-increasing hunting pressure. Numbers for western gorillas are more likely to fall in the tens of thousands, and for the eastern lowland gorillas of DRC, in the thousands.

As if hunting weren't bad enough, another catastrophic threat has recently emerged:

Ebola. This threatens not just gorillas, but chimpanzees as well. For instance, in a gorilla sanctuary in Republic of Congo, Ebola virus was confirmed in several gorilla and chimpanzee carcasses found by researchers in November and December 2002. A survey of the forest indicated that by February 2003, there had been a massive die-off of apes from the reserve, including all eight gorilla groups that had been monitored by researchers since 1994.

Hopeless as the situation seems, the news isn't all bad. The last 15 years have seen a proliferation of conservation efforts across Africa. People have taken the Mountain Gorilla Project as a blueprint and adapted it for use in other regions. There are now long-term conservation programs in place in every country that harbors gorillas, and in west-central Africa, at least six national parks have been created with gorillas as a focus.

The Future

Gorilla conservation encompasses myriad activities and initiatives, from research, to regulating tourism, to involving local human communities in conservation work. Generally speaking, however, the two most crucial tasks for the future are, firstly, establishing and securing protected areas. The only hope for wildlife in Africa is to

Caretakers help rehabilitate orphans in Gabon.

guarantee a place for it to live, and the only way to do this is to secure the borders of reserves and national parks. The other critical task is controlling the bushmeat trade. Campaigns to fight this trade have gained momentum, and include far-reaching initiatives aimed at foreign timber companies that facilitate the killing and transport of wild meat.

These tasks are daunting, but there are glimmers of hope. When I get discouraged about the future of gorillas and other wildlife, I take heart from what happened in the Virunga Volcanoes.

The 1990s were a horrific time for the entire region. Rwanda suffered a war and one of the worst genocides in human history. A flood of half a million refugees poured through the gorillas' habitat, fleeing into DRC. Land mines dotted the edge of the National Park and the forest still remains a hiding place for rebels.

Karisoke Research Center was completely destroyed. Nevertheless, through it all, habituated gorilla groups have been monitored nearly continuously by park rangers, conservation workers and researchers. The perseverance and courage of these people in both DRC and Rwanda, in the face of seemingly impossible conditions, is truly inspirational.

A recent census showed that, so far, the gorillas of the Virunga Volcanoes have survived and even thrived during the last ten years. If the animals can persist in the face of such odds, it gives hope for other regions.

Is it more important to save gorillas and the other great apes than it is to save other species? On the practical side, to protect apes, you must protect large tracts of natural tropical forest, which means you will be preserving all the other fauna and flora that exist there. But there is an argument that says we have a greater moral obligation to the great apes than to other species because of their evolutionary closeness to ourselves. Some people see this attitude as typical human arrogance and chauvinism. Perhaps. But there is no doubt that when you sit with a group of wild gorillas, you feel a connection. The way they move their hands, the expressions in their faces, seem at times intimately familiar.

If we let gorillas and the other great apes slip away, we will have lost our closest connection to our past, to our biological heritage. We will have lost forever a chance to understand this past and ourselves more fully. We will have lost our closest link to the rest of life on earth.

Distribution of the World's Gorillas

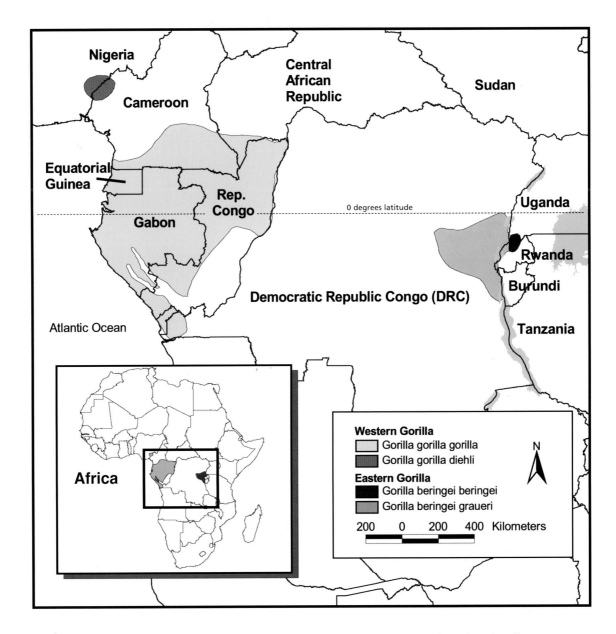

The shaded areas outline the regions within which the different species and subspecies of gorillas occur.
Distribution of the animals within these shaded areas is not continuous.

Gorillas Fact File

Species subspecies	Western gorilla *Gorilla gorilla* western lowland *G.g. gorilla* Nigerian *G.g. diehli*
Species subspecies	Eastern gorilla *Gorilla beringei* mountain *G.b. beringei* eastern lowland *G.b. graueri*
Weight	Adult males 308–352 lb (140–160 kg) Adult females 154–176 lb (70–80 kg)
Gestation	256 days (8½ months)
Birth weight	4.6 lb (2.1 kg)
Age of weaning	2.5–3 years
Birth interval	4 years
Male age at full maturity	13–15 years
Female age at first birth	10 years
Longevity (eastern gorillas)	Males 25–30 years Females 28–33 years
Average group size	10
Yearly range size	Western and eastern lowland gorilla 4–12 sq miles (11–31 sq km) Mountain gorilla 3 sq miles (8 sq km)

Data shown are approximate or average values for wild gorillas. Weights are the ranges for both species. 'Longevity' is the usual age of death for animals once they have survived into adulthood, and refers only to eastern gorillas, as there are too few data from west-central Africa to make an estimate. Figures for yearly range size of western and eastern lowland gorillas are average range sizes across several study sites.

Recommended Reading

George Schaller's *The Mountain Gorilla* (1963, Chicago University Press), and his more popularly written, *The Year of the Gorilla* are excellent accounts of the first intensive study of the behavior and ecology of wild mountain gorillas, and provide a wealth of information on many aspects of gorillas' lives.

Gorillas in the Mist by Dian Fossey (1983, Hodder and Stoughton) is a first-person account of Fossey's work with mountain gorillas.

In the Kingdom of Gorillas by Bill Weber and Amy Vedder (2001, Simon and Schuster) was written for the interested public and chronicles the establishment and running of conservation projects focused on mountain gorillas.

For a synthesis of scientific studies on gorillas, *Mountain Gorillas* (2001, Robbins, M., Sicotte, P. and Stewart, K.J., eds. Cambridge University Press) reviews studies from over 30 years of research on mountain gorillas, and includes papers on the other subspecies for comparison.

Gorilla Biology (2003, Taylor, A.B. and Goldsmith, M.L., eds. Cambridge University Press) presents recent papers on all subspecies of gorillas, covering a wide variety of topics, from genetics to conservation.

Index

adolescence, adolescents 24, 27, 53
aggression 9, 20, 39
alpha male 40, 45, 48
bachelors 53, 54
bachelor groups 54
birth 7, 23, 24, 27, 35, 42, 71
blackbacks 24, 48
brachiation 17
breeding 35, 45, 46, 48, 49, 53
bushmeat 64, 67
census 60, 68
charges, charging 7, 8, 9, 49
chest-beating 17, 18, 36, 49
chimpanzees 8, 13, 17, 21, 33, 67
Congo 10, 54, 56, 59
conservation 59, 60, 63, 67, 68
death 28, 35, 40, 42, 50, 53
Democratic Republic of Congo
    (DRC) 9, 10, 14, 54, 66, 68
diet 18, 21, 31, 32
disease 28, 60, 63
disperse, dispersal 42, 46, 48, 54, 59
disturbance 60
DNA 14, 48
dominance 40, 46, 48, 49
    rank 40, 48
Du Chaillu 9
eastern gorillas 14, 42, 71
eastern lowland gorillas 14, 42, 54,
    56, 66, 71
Ebola 28, 67

emigration 35, 42, 45
estrus 40
fighting, fights 18, 20, 28, 49, 50
fission 33, 42
foliage 31, 32, 33, 36
folivory 32, 33
Fossey, Dian 10, 49, 59, 72
fossils 13
fruit 31, 32, 33, 42
Gabon 8, 64, 66
genocide 68
groups 9, 10, 15, 18, 20, 21, 27, 31,
    32, 33, 35, 36, 39, 40, 42, 45, 46,
    48, 49, 50, 53, 54, 56, 59, 60, 63,
    68, 71
growth 23, 24
habitat 10, 14, 32, 33, 59, 63, 64, 68
habituation 36, 54, 56, 59, 60, 63,
    68
home range 31, 32
hunters, hunting 9, 33, 64, 66
immigration 35, 42
inbreeding 45, 46
infancy, infants 7, 20, 23, 24, 27, 28,
    39, 46, 50, 53, 56, 59, 64
infanticide 50, 53, 54, 56, 59
IUCN 64
Kahuzi-Biega National Park 10, 54,
    56, 66
Karisoke Research Center 10, 24,
    35, 36, 42, 63, 68

knuckle-walking 17
lactation 27, 50
language 21
life expectancy 28
logging 64
mating 10, 20, 24, 27, 45, 46, 48, 49,
    50, 53, 54
Mountain Gorilla Project (MGP)
    59, 60, 63, 64, 67
mountain gorillas 10, 14, 24, 27, 28,
    31, 32, 35, 36, 40, 45, 46, 53, 54,
    56, 60, 63, 64, 72
multimale groups 35, 48, 53
natal group 45, 46, 48
national parks 9, 10, 54, 56, 59, 60,
    64, 66, 67, 68
nests 24, 39, 56
orangutans 8, 13, 39
orphan 39
poachers, poaching 56, 59, 60, 64
polygyny 35
predators 20, 28, 40
pregnancy 27
primates 10, 13, 17, 39, 50
range 9, 49, 64
reproduction 24
Republic of Congo 66, 67
rest period 21, 39, 50
Rwanda 9, 10, 14, 59, 60, 63, 68
Savage, Thomas 8
Schaller, George 10, 72

sexual dimorphism 18, 36
silverbacks 7, 8, 9, 18, 28, 35, 36, 39,
    40, 42, 46, 48, 49, 50, 53, 56, 59,
    64
smell, sense of 20
social
    behavior 10, 33, 35, 39
    relationships 36, 42
    standing 40
    world 23
sociality 31, 32
solitary males 35, 49, 50, 53, 54, 56
stature 18
subspecies 10, 14, 31, 54, 72
swamps 33
termites 21, 33
threat 10, 40, 64, 66
    displays 18, 49
tools 21
tourism 59, 60, 63, 64, 67
transfer 46, 54, 56
Uganda 9, 10, 14, 60, 63
Virunga Volcanoes 9, 10, 27, 31, 32,
    35, 36, 42, 50, 54, 60, 63, 64, 68
vocalization 20, 21, 27, 40
war 56, 63, 66, 68
weight 17, 18, 23, 71
west-central Africa 28, 31, 33, 42,
    63, 67
western gorillas 10, 14, 32, 33, 35,
    42, 54, 63, 64, 66, 71

Biographical Note

Kelly Stewart is a Research Associate in the Anthropology Department at the University of California in Davis. She first went to Rwanda in 1973 to be research assistant to Dian Fossey at the Karisoke Research Center, and then did her dissertation on the behavior of immature gorillas, receiving her degree from Cambridge University in 1982.

With her husband, Alexander Harcourt, Stewart co-directed Karisoke Research Center from 1981-1983, and has also worked in Nigeria, conducting a census of western gorillas. She edited the annual *Gorilla Conservation News* for 10 years, has written numerous scientific and popular articles on the behavior, ecology and conservation of mountain gorillas, and was an editor of a recent book summarizing more than three decades of research from Karisoke Research Center.